A Beginning-to-Read Book

LIFE SCIENCE

WHERE PLANTS LIVE

by Mary Lindeen

NorwoodHouse Press

DEAR CAREGIVER, The *Beginning to Read—Read and Discover Science* books provide young readers the opportunity to learn about scientific concepts while simultaneously building early reading skills. Each title corresponds to three of the key domains within the Next Generation Science Standards (NGSS): physical sciences, life sciences, and earth and space sciences.

The NGSS include standards that are comprised of three dimensions: Crosscutting Concepts, Science and Engineering Practices, and Disciplinary Core Ideas. The texts within the *Read and Discover Science* series focus primarily upon the Disciplinary Core Ideas and Crosscutting Concepts—helping readers view their world through a scientific lens. They pique a young reader's curiosity and encourage them to inquire and explore. The Connecting Concepts section at the back of each book offers resources to continue that exploration. The reinforcement activities at the back of the book support Science and Engineering Practices—to understand how scientists investigate phenomena in that world.

These easy-to-read informational texts make the scientific concepts accessible to young readers and prompt them to consider the role of science in their world. On one hand, these titles can develop background knowledge for exploring new topics. Alternately, they can be used to investigate, explain, and expand the findings of one's own inquiry. As you read with your child, encourage her or him to "observe"—taking notice of the images and information to formulate both questions and responses about what, how, and why something is happening.

Above all, the most important part of the reading experience is to have fun and enjoy it!

Sincerely,

Shannon Cannon

Shannon Cannon, PhD
Literacy Consultant

Norwood House Press
For more information about Norwood House Press please visit our website at www.norwoodhousepress.com or call 866-565-2900.

Editor: Judy Kentor Schmauss **Designer**: Sara Radka

Photo Credits: Getty Images: cover, 4, 6, 7, 9, 10, 12, 13, 16, 17, 18, 19, 22, 26, 28, 29; Shutterstock: 1, 3, 8, 12, 14, 15, 20, 24, 29

Library of Congress Cataloging-in-Publication Data
Names: Lindeen, Mary, author.
Title: Where plants live / Mary Lindeen.
Description: Chicago : Norwood House Press, [2022] | Series: Beginning-to-read | Audience: Grades K-1 | Summary: "Earth is home to millions of plants. Featured plants include those from forest, desert, mountain, rain forest, prairie, freshwater, and saltwater habitats. Includes science and reading activities and a word list"— Provided by publisher.
Identifiers: LCCN 2021019800 (print) | LCCN 2021019801 (ebook) | ISBN 9781684508273 (hardcover) | ISBN 9781684046553 (paperback) | ISBN 9781684046614 (epub)
Subjects: LCSH: Plants—Habitat—Juvenile literature.
Classification: LCC QK901 .L56 2022 (print) | LCC QK901 (ebook) | DDC 581.7—dc23
LC record available at https://lccn.loc.gov/2021019800
LC ebook record available at https://lccn.loc.gov/2021019801

Hardcover ISBN: 978-1-68450-827-3 Paperback ISBN: 978-1-68404-655-3

PO339N—082021
Manufactured in the United States of America in North Mankato, Minnesota.

Plants need water, warmth, food, and sunlight in order to grow.

Plants can only live where they can get what they need.

These plants can grow in a forest.

Plants like these need a lot of sunlight and warmth but not much water.

They can grow in the desert.

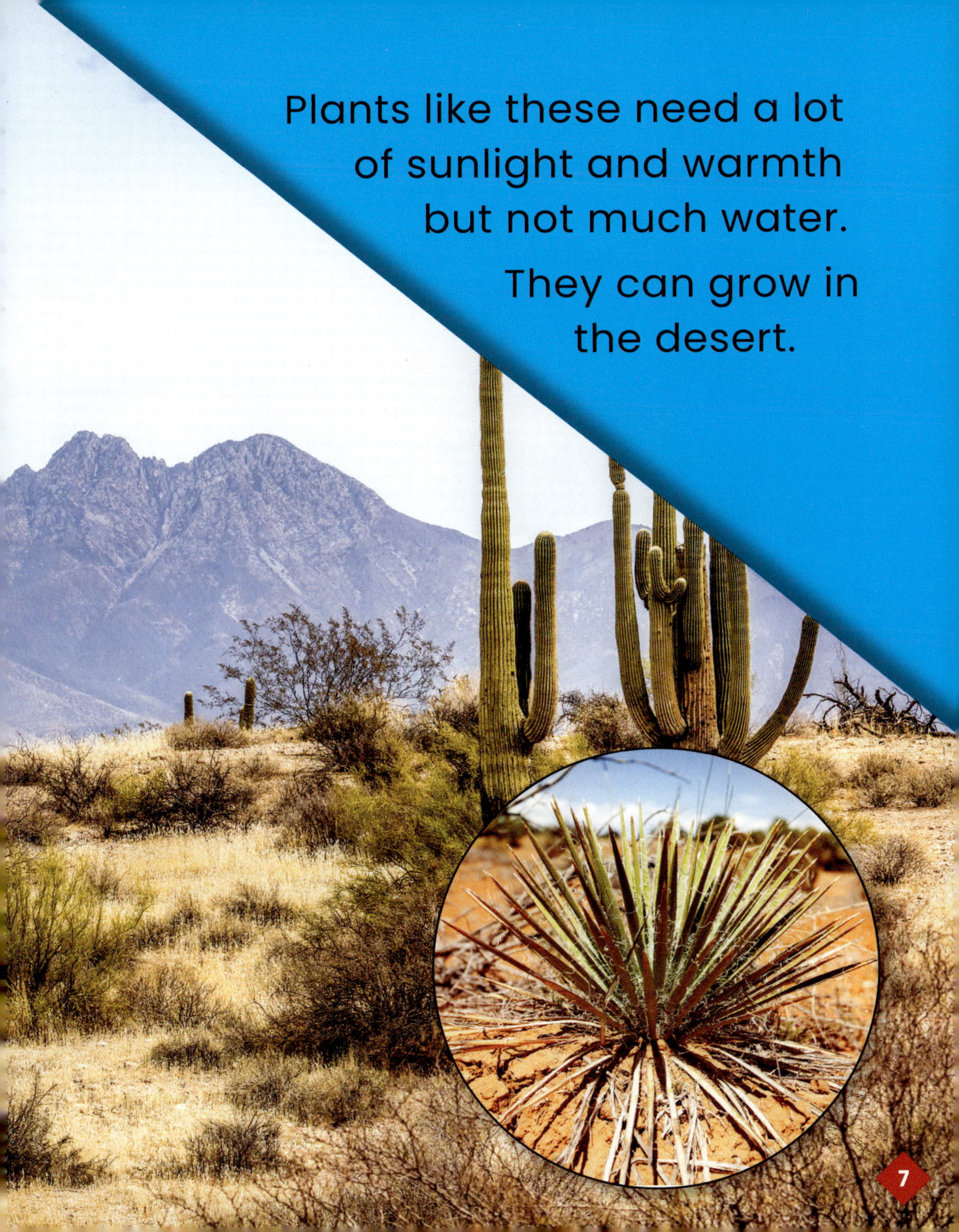

Plants like these can live where it is rocky and cold. They can grow in the mountains.

Did You Know?

Some rain forest trees have very smooth bark. It helps the trees shed water and makes it hard for bugs to crawl on them.

Rain forest plants need warm weather and lots of rain.

Prairie plants need lots of sun. They can grow where it's hot in the summer and cold in the winter.

Some plants need to live in water to get what they need.

These plants grow in fresh water.

Other plants need salt water.

They grow in the ocean.

Some plants grow in shallow water.

Other ocean plants
float in deep water.

Many plants grow where nature puts them.

Other plants grow where people put them.

Some plants grow in small spaces.

But other plants need a lot of room to grow.

Did You Know?

Banyan trees grow from seeds that land on other trees. Banyan roots look like new tree trunks.

And some plants can even grow in your home!

Plants That Grow on Land

Plants That Grow in Water

... READING REINFORCEMENT...

CONNECTING CONCEPTS

CLOSE READING OF NONFICTION TEXT

Close reading helps children comprehend text. It includes reading a text, discussing it with others, and answering questions about it. Use these questions to discuss this book with your child:

- What are three things plants need to grow?
- Where do plants grow if they do not need much water?
- Do saltwater plants grow in lakes or oceans?
- Where do plants grow if they can survive in hot weather and cold weather?
- What did you learn about plants from this book that you didn't know before?

SCIENCE IN THE REAL WORLD

Ask your child to identify the kind of place you live in, such as a desert, a prairie, the mountains, and so on. Talk about the plants that grow there, why they grow there, and what they need to survive.

SCIENCE AND ACADEMIC LANGUAGE

Make sure your child understands the meaning of the following words:

warmth **forest** **desert** **rock** **rain forest**

fresh water **salt water** **shallow** **nature**

Have him or her use the words in a sentence.

FLUENCY

Fluency is the ability to read accurately with speed and expression. Help your child practice fluency by using one or more of the following activities:

1. Reread the book to your child at least two times while he or she uses a finger to track each word as it is read.

2. Read a line of the book, then reread it as your child reads along with you.

3. Ask your child to go back through the book and read the words he or she knows.

4. Have your child practice reading the book several times to improve accuracy, rate, and expression.

FOR FURTHER INFORMATION

Books:

Cassany, Mia. *Wilderness: Earth's Amazing Habitats.* Munich: Prestel Junior, 2019.

Stewart, Melissa. *A Seed Is the Start*. Washington, DC: National Geographic, 2017.

Williams, Kathryn. *Plants*. Washington, DC: National Geographic, 2017.

Websites:

National Geographic Kids: Habitats
https://kids.nationalgeographic.com/nature/habitats

SmartClass 4 Kids: Plant Adaptations
https://smartclass4kids.com/science/plants-facts/plant-adaptation/

Text Project: Plants and Where Plants Grow
http://textproject.org/teachers/vocabulary-instruction/textproject-word-pictures/core-vocabulary/plants-and-where-plants-grow/

Word List

Where Plants Live uses the 85 words listed below. *High-frequency* words are those words that are used most often in the English language. They are sometimes referred to as sight words because children need to learn to recognize them automatically when they read. *Content* words are any words specific to a particular topic. Regular practice reading these words will enhance your child's ability to read with greater fluency and comprehension.

HIGH-FREQUENCY WORDS

a	have	make(s)	other	these
and	help(s)	many	people	they
but	home	much	put(s)	to
can	in	new	small	very
even	is	not	some	water
for	it	of	that	what
from	like	on	the	where
get	look	only	them	your

CONTENT WORDS

banyan	forest	mountains	rocky	summer
bark	fresh	nature	room	sun
bugs	grow	need	roots	sunlight
cold	hard	ocean	salt	tree(s)
crawl	hot	order	seeds	trunks
deep	it's	plants	shallow	warm
desert	land	prairie	shed	warmth
float	live	rain	smooth	weather
food	lot(s)	rain forest	spaces	winter

About the Author

Mary Lindeen is a writer, editor, parent, and former elementary school teacher. She has written more than 100 books for children and edited many more. She specializes in early literacy instruction and books for young readers, especially nonfiction.

About the Advisor

Dr. Shannon Cannon is an elementary school teacher in Sacramento, California. She has served as a teacher educator in the School of Education at UC Davis, where she also earned her PhD in Language, Literacy, and Culture. As a member of the clinical faculty, she supervised pre-service teachers and taught elementary methods courses in reading, effective teaching, and teacher action research.